Florida Citrus Varieties

D. P. H. Tucker
S. H. Futch
F. G. Gmitter
M. C. Kesinger

UNIVERSITY OF FLORIDA
Institute of Food and Agricultural Sciences

Florida Citrus Varieties

D. P. H. Tucker, S.H. Futch, F.G. Gmitter, M.C. Kesinger

D. P. H. Tucker is a Professor-Extension Horticulturist, CREC-Lake Alfred; S. H. Futch is an Extension Agent-Multi-County, CREC-Lake Alfred; F. G. Gmitter is a Horticulturist, CREC-Lake Alfred; and M. C. Kesinger, Chief, Bureau of Citrus Budwood Registration, DPI, FDACS.

Edited by James Mathis
Design and Layout by Helen Huseman
Logo by Yaeko Egashira

First Printing: August 1982
Second Printing: October 1986
Revised: January 1993
Revised: September 1998

© Copyright 1998, University of Florida, Institute of Food and Agricultural Sciences. All rights reserved. Parts of this publication may be reproduced for educational use. Please provide credit to the "University of Florida, Institute of Food and Agricultural Sciences."

Second Edition

Cover photo and photos in the book were kindly provided by the Florida Department of Citrus.

Related titles of interest:

Florida Citrus CD-Rom (SW 120), $50
Florida Citrus Rootstock Selection Guide (SP 248), $20
Identification of Mites, Insects, Diseases, Nutrition, Symptoms and Disorders of Citrus (SF 07), $10
Your Florida Dooryard Citrus Guide (SP 178), $5

For ordering information please call 1-800-226-1764 or write:

University of Florida/IFAS Publications
PO Box 110011
Gainesville, FL 32611-0011

Contents

INTRODUCTION ... 1
FLORIDA CITRUS MATURITY PERIODS ... 3

ORANGES
Navel ... 6
Cara Cara ... 7
Ambersweet ... 8
Hamlin ... 9
Parson Brown ... 10
Pineapple ... 11
Sunstar ... 12
Midsweet ... 13
Gardner ... 14
Valencia ... 15
Sanguinelli ... 16

GRAPEFRUIT
Duncan ... 19
Marsh ... 20
Foster ... 21
Thompson ... 22
Redblush ... 23
Ray Ruby ... 24
Flame ... 25
Rio Red ... 26
Star Ruby ... 27

TANGERINES AND TANGERINE HYBRIDS
Minneola ... 30
Orlando ... 31
Nova ... 32
Robinson ... 33
Sunburst ... 34
Dancy ... 35
Murcott ... 36
Temple ... 37
Osceola ... 38
Lee ... 39
Fallglo ... 40
Page ... 41
Satsuma ... 42
Ponkan ... 43

ACID CITRUS FRUIT
Tahiti (Lime) ... 46
Key (Lime) ... 47
Meyer (Lemon) ... 48
Bearss (Lemon) ... 49

MISCELLANEOUS
Calamondin ... 52
Nagami (Kumquat) ... 53
Meiwa (Kumquat) ... 54
Tavares (Limequat) ... 55
HiradoBuntan Seedling (Pummelo) ... 56
Nakon (Pummelo) ... 57

ACKNOWLEDGEMENTS

The current authors wish to acknowledge the contributions of C. J. Hearn and C. O. Youtsey (both now retired) as coauthors of the original publication.

The authors also express appreciation to Randall C. Smith, United States Department of Agriculture (USDA), Agricultural Research Service (ARS), for assistance in preparing photographs used in this publication.

Introduction

Florida produces a wide selection of high quality citrus fruits for processing into juice products and for fresh consumption. For a large tree fruit industry, such as citrus in Florida, where over 85 percent of the product is processed, it is important to retain a viable, well identified fresh fruit sector which enjoys a high degree of recognition by the general public in the United States and abroad.

The purpose of this publication is to provide brief descriptions of citrus cultivars suitable for planting under Florida conditions so that commercial growers, homeowners and other interested individuals can make informed decisions based on their respective needs. Suitable selections from the range of varieties described will provide fruit for sale or personal consumption throughout much of the year. The text identifies origins of the various cultivars and provides information on fruit size, seediness and maturity periods as well as horticultural and pest-related characteristics of the fruit and tree.

Commercial producers, nurserymen, packers, processors, retail outlets and homeowners are beneficiaries of the many horticultural, plant protection, postharvest and marketing standards developed over the years. The Citrus Budwood Protection Program administered by the Florida Department of Agriculture and Consumer Services (FDACS), Division of Plant Industry (DPI), Budwood Registration Bureau evaluates promising citrus varieties for bud-transmissible diseases and horticultural characteristics, including yield and fruit quality, and makes propagation material from selected trees available to nurseries which produce trees for the industry and retail outlets. Current information on numbered selections (clones or registered budlines) of varieties is available from the Bureau. DPI also maintains an arboretum located in Winter Haven, Florida where many of the varieties in this publication can be seen. Regulations promulgated by the Department are intended for the welfare of the commercial industry, consumers and the economy of the State of Florida.

Selection of dooryard citrus trees should also be based on standards similar to those considered by commercial growers, namely end use (utilization), maturity period, tree size, cold, flood salt tolerance, other soil-related factors such as pH and pest and disease resistance. Fruit size, eating quality, seediness, ease of peeling and on-tree holding quality are also of interest to commercial producers, dooryard growers, and consumers.

For citrus to be a profitable venture there must be a willingness to provide a high level of management input and to become very knowledgeable about specific cultural, harvesting, handling and marketing requirements of individual varieties for the fresh and juice products markets. Management of tree growth, fruit production and quality integrates site characteristics, predominantly soil type, variety, rootstock vigor, tree spacing, nutritional, water and weed control inputs and eventually hedging and topping. The objective is to optimize fruit production and quality with minimal vegetative regrowth. Citrus rootstocks affect more than 20 horticultural and pathological characteristics of the tree and fruit including tree vigor, cold hardiness, yield, fruit size, juice quality and holding quality on the tree and pest tolerance.

Trees of different varieties vary in degree of cold hardiness with mandarin types being the most hardy, Temple and Fallglo being the notable exceptions. Sweet oranges rank next in cold hardiness, closely followed by grapefruit. Lemons and limes (acid fruit) are far less cold hardy and are, therefore, limited to the warmest locations. Fruit tolerance to cold differs from tree hardiness, with grapefruit usually being the most tolerant due to fruit size and peel thickness, followed by oranges, mandarins, lemons and limes.

Cross pollination is necessary for satisfactory fruit set and yields of certain mandarin types and hybrids. The pollenizer variety, its percentage of the planting and arrangement and its marketing potential are all very important. Growers should not overlook the importance of honeybees and their behavior for effective cross pollination. Fertilization and irrigation practices also influence fruit size, internal and external quality and sometimes storage and shipping qualities. Over fertilization, over irrigation and frequent heavy pruning invariably results in excessive vegetative growth, poorer fruit quality and in some cases greater susceptibility to foliar fungal diseases. Excessive crop load and fruit size, sometimes associated with limb breakage in certain varieties such as Murcott, Dancy and Sunburst, can be regulated by mechanical pruning or other fruit thinning practices.

A higher level of pest and disease management is required for fresh fruit as external appearance is important for the fresh fruit market. In addition to wind scar, considerable surface blemish can result from mites, insect and disease infestations in Florida. Cultural practices may be used to modify the impact of pests and diseases but agrichemicals will have to be used if monitoring indicates a necessity. While the commercial fresh market demands blemish-free fruit, the homeowner who is usually more interested in the eating quality may choose not to spray repeatedly to maintain external appearance. Reduced frequency of overhead irrigation or its replacement by low volume, subcanopy systems will reduce the incidence and severity of some fungal diseases.

In the absence of some fungicides previously available it would be wise to select varieties not highly susceptible to the fungus diseases such as scab and *Alternaria* brown spot, which require multiple sprays for their effective control.

Citrus matures slowly, does not suddenly ripen as do many other fruit and does not continue to ripen after harvest. Gradual changes in juice content, sugar and acid levels determine fruit maturity, with acid content decreasing and sugar content increasing as fruit mature. Commercial growers take samples of fruit for analysis to determine if legal maturity standards have been attained. The dooryard grower usually determines this by fruit color break and taste. Citrus may be held on the tree long after it attains acceptable quality standards, usually improving in taste until the flesh begins to dry out, at which time it may fall from the tree. Frozen fruit will drop within a few days and those remaining on the tree will dry out with time, with the degree of juice loss being related to variety and the severity and duration of freezing temperatures. The homeowner may pick such fruit and store the juice in the freezer. Slightly damaged Valencia fruit will often recover some of their juice content with a relatively early freeze. A grapefruit crop from the same bloom can be harvested from October through May or later, but fruit drop and seed germination in the fruit may occur later in the season. Fruit of orange varieties usually hold for shorter periods than grapefruit, while mandarin types cannot be stored on the tree for nearly as long. Mandarins tend to become puffy with age when the flesh separates from the peel. Lemons and limes may be used whenever deemed to have enough juice. Although the major citrus bloom is in the spring, acid varieties will bloom sporadically making it possible to have a limited quantity of fruit much of the year.

The orange peel color of fruit, which is closely identified with eating quality by consumers, is intensified by cool fall and winter temperatures. Thus, fruit sometimes will not develop the desired color in Florida and in tropical areas with milder climate conditions. Flesh color is similarly affected but to a lesser degree. Early maturing varieties generally are not as well colored as mid- and late-season ones and the better colored later maturing Valencia may actually regreen to some extent if held on the tree until late in the season. Fruit of mandarin types, tangerines and tangelos, vary widely in color, some being more dependent on low temperature for color development than others. Grapefruit develops an excellent peel color even in the hottest climates, but the green color changes to yellow earliest in cooler climates. Fruit shape and peel texture, important factors in fresh fruit marketing, are also qualities influenced by climate. Lemons are picked according to size and degreened with ethylene to develop the yellow color. Limes are picked for juice content and size, are normally green when "ripe" but do eventually become yellow.

As much time as possible should be allowed for fruit to reach peak maturity on the tree, as quality usually improves with time. More emphasis should be placed on the sustained production of high quality, marketable fruit than on early returns from fresh fruit of lesser quality harvested and marketed too early. A consumer is less likely to be a return customer following an unsatisfactory experience. Fruit must be handled carefully during harvesting and handling (including degreening operations) to minimize fruit damage and postharvest decay development, and to improve shelf life. Certain varieties attain internal quality standards before acceptable external color development, while others dry out prematurely and, therefore, require early harvesting. Where fruit is prepared for sale on the fresh market, Florida Department of Citrus regulations should be followed and postharvest fungicidal and wax treatments properly applied. If fruit is stored, correct conditions of relative humidity and temperature should be maintained.

Fresh fruit will usually, but not always, return higher prices than fruit for processing and is, therefore, a good hedge against years of low processed product returns. Certain fresh citrus fruit varieties may be used for juice blending to improve juice color, an increasingly important marketing factor. Valencias held late in the season may command a higher price for the not from concentrate (NFC) market.

The United States Department of Agriculture and University of Florida citrus fruit breeding programs are integrating emerging biotechnologies that will facilitate the development and selection of improved fresh and processing varieties that: 1) satisfy demands between marketing periods of existing varieties and capture early markets with varieties which may be harvested before the threat of freezing weather; 2) improve external appearance (particularly fruit color), ease of peeling, internal quality, juice color and reduce seed content; 3) improve storage, shipping qualities and shelf life; 4) reduce pest and disease susceptibility of tree and fruit, thereby lowering dependence on agrichemicals; and 5) improve cold hardiness of trees to better survive freezes. Emphasis is on the development of high quality seedless mandarin types for the fresh market and improved sweet orange selections that will facilitate the processing industry's shift from frozen concentrate to not from concentrate (NFC).

Florida Citrus Harvesting Periods*

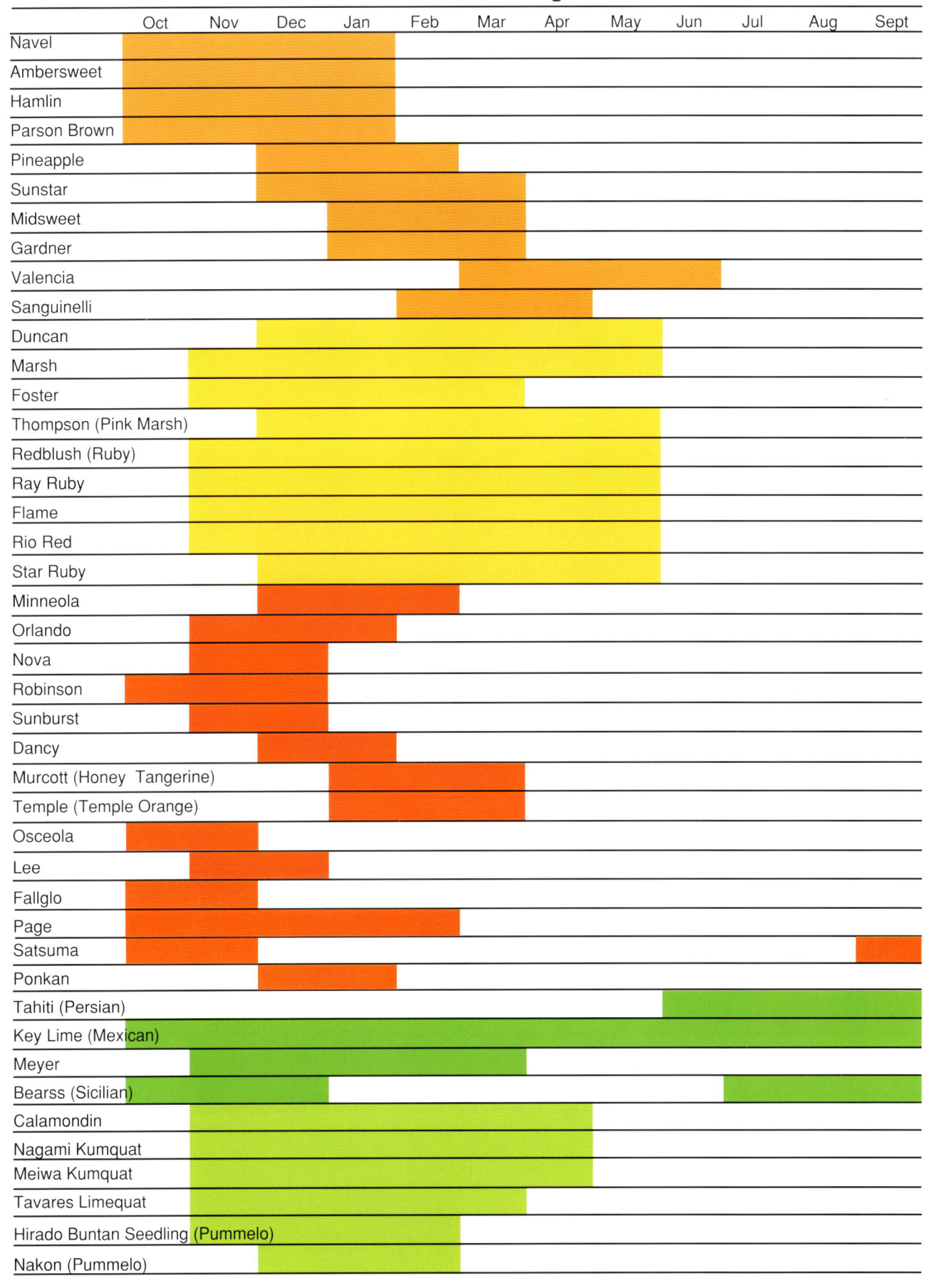

Variety	Oct	Nov	Dec	Jan	Feb	Mar	Apr	May	Jun	Jul	Aug	Sept
Navel	●	●	●	●								
Ambersweet	●	●	●	●								
Hamlin	●	●	●	●								
Parson Brown	●	●	●	●								
Pineapple			●	●	●	●						
Sunstar			●	●	●	●						
Midsweet				●	●	●						
Gardner				●	●	●						
Valencia						●	●	●	●			
Sanguinelli					●	●	●					
Duncan			●	●	●	●	●	●	●			
Marsh			●	●	●	●	●	●	●			
Foster			●	●	●	●	●	●	●			
Thompson (Pink Marsh)			●	●	●	●	●	●	●			
Redblush (Ruby)			●	●	●	●	●	●	●			
Ray Ruby			●	●	●	●	●	●	●			
Flame			●	●	●	●	●	●	●			
Rio Red			●	●	●	●	●	●	●			
Star Ruby			●	●	●	●	●	●	●			
Minneola			●	●	●	●						
Orlando		●	●	●								
Nova		●	●									
Robinson	●	●	●									
Sunburst		●	●	●								
Dancy			●	●								
Murcott (Honey Tangerine)					●	●	●					
Temple (Temple Orange)				●	●	●						
Osceola	●	●										
Lee	●	●	●									
Fallglo	●	●										
Page		●	●	●	●							
Satsuma												●
Ponkan			●	●								
Tahiti (Persian)									●	●	●	
Key Lime (Mexican)			●	●	●	●	●	●	●	●	●	●
Meyer			●	●	●							
Bearss (Sicilian)			●	●	●				●	●	●	
Calamondin			●	●	●	●	●	●				
Nagami Kumquat			●	●	●	●	●	●				
Meiwa Kumquat			●	●	●	●	●	●				
Tavares Limequat			●	●	●	●	●	●				
Hirado Buntan Seedling (Pummelo)			●	●	●	●						
Nakon (Pummelo)			●	●	●	●						

*Maturity period of each variety actually exceeds range shown as maturity is affected by Florida's seasonal and production location variations. Harvest of grapefruit in southern Florida begins usually in late September-early October with legal maturity obtained 2-3 weeks later in central and northern growing areas. Best eating quality for grapefruit is usually not attained until November.

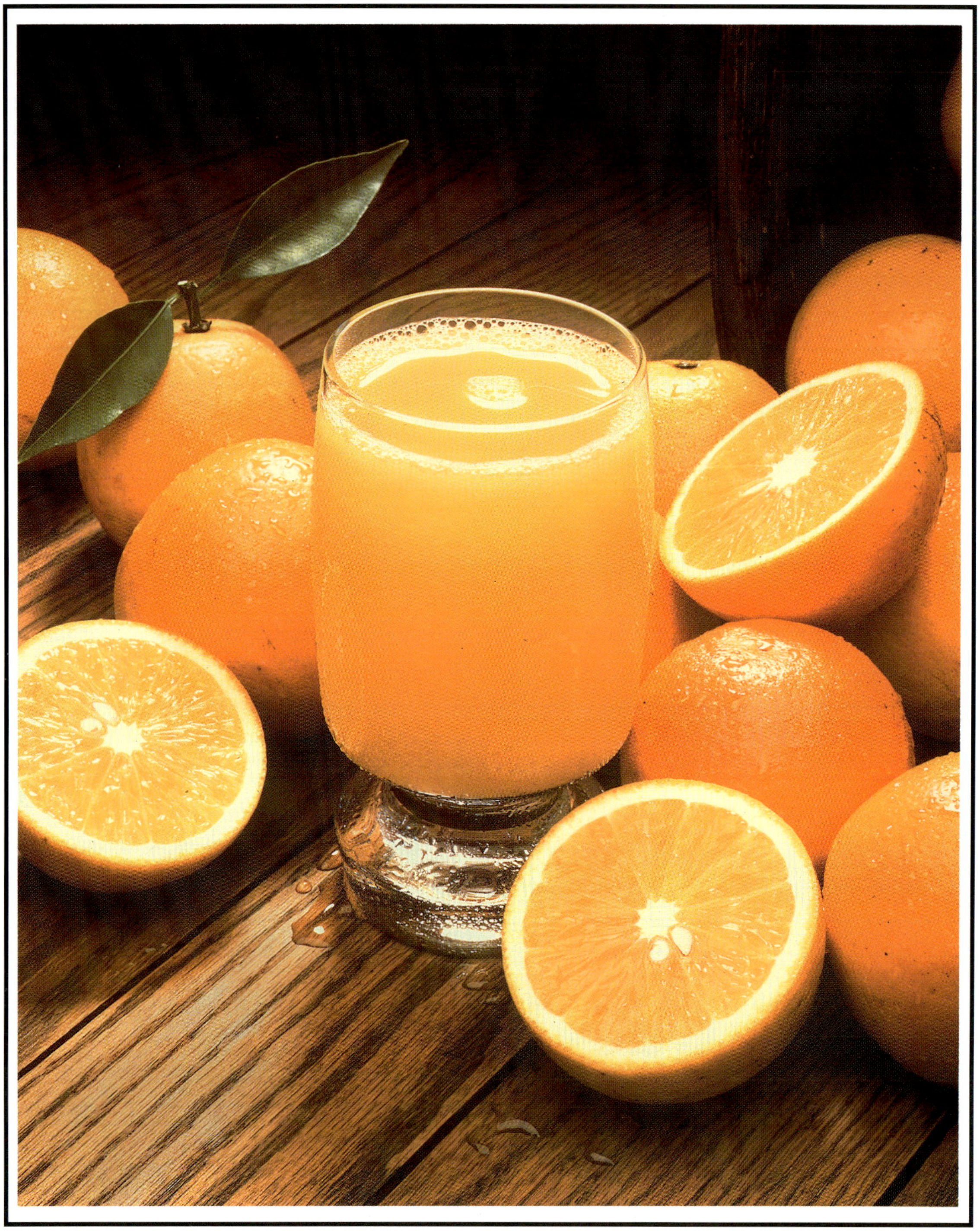

ORANGES

The sweet orange was cultivated in China before being introduced to Europe during the fifteenth century. Columbus is credited with bringing sweet orange seeds to the New World during his second voyage in 1493. The first plantings of sweet orange in the United States were established in Florida between 1513 and 1565 in and around the settlement of St. Augustine. The sweet orange was spread rapidly throughout Florida by Spanish explorers and Indians. Sweet orange trees are moderately large, reaching a height of 30 feet or higher and 20 feet or more in diameter if widely spaced and left unpruned. The fruit of sweet orange trees and all other citrus is classified as a specialized berry known as a hesperidium. Fruit is spherical to oblong and usually 2.5 to 4.0 inches in diameter, but will vary with variety and crop load. The seed number is variable, with some varieties having zero to six seeds (commercially seedless), while others contain as many as 15 to 20 seeds. Sweet orange varieties can be categorized into four distinct groups: round oranges, navel oranges, blood or pigmented oranges, and acidless oranges.

Variety:
NAVEL

Type and parentage: Sweet orange
Average diameter (inches): 3 - 3 1/2
Seeds per fruit: 0 - 6
Commercial harvest season:
October - January

The navel orange, most likely grown in the Mediterranean region prior to its introduction to Brazil by the Portuguese, was introduced to the United States in 1870. Navel oranges are well suited to Mediterranean and subtropical climates where they attain good external and internal color. Extreme tropical climates are not well suited for navel production which limits adaptability worldwide. Navel oranges are characterized by a small, secondary fruit embedded in the stylar end of the primary fruit. The size and appearance of the secondary fruit varies with the variety, being inconspicuous in some while quite prominent in others. Fruit are typically large, commercially seedless and early ripening. They have a distinctive flavor, are easily peeled and sectioned and tend to be lower in acid content than most orange varieties. Navel oranges tend to be genetically unstable so careful selection is necessary to maintain true-to-type propagations. The most popular variety is 'Washington' navel, but a number of other selections are available. Selections available in Florida produce relatively small crops of large fruit. Navel oranges are primarily used for fresh consumption rather than processing, because delayed bitterness due to limonin often develops making the juice unpalatable. When used for juice, they must be blended with other varieties with low limonin content. Navel oranges are susceptible to environmental stresses and physiological disorders which often result in low yields. Poor fruit set, postbloom fruit drop (PFD) caused by the fungus *Colletotrichum acutatum*, and several fruit drop events reduce fruit yields. Fruit drop periods during early and late summer, which account for 15-20% of the crop in some years, may be alleviated with growth regulator sprays. Fruit from trees on more vigorous rootstocks tend to be excessively large and dry out prematurely. Rind blemishes (oleocellosis) can occur early in the season from rough handling at harvest. The fruit tends to split and also develop fungal infection at the blossom end, particularly where there is a pronounced opening. Navels tend to require more precise management, particularly relating to stress avoidance, and irrigation and nutrition levels should be carefully monitored.

Variety:
CARA CARA
(RED NAVEL)

Type and parentage:
Sweet orange
Average diameter (inches): 3 - 3 1/2
Seeds per fruit: 0 - 6
Commercial harvest season:
October - January

The Cara Cara red navel was introduced from Venezuela and released for propagation in Florida in 1990. The Cara Cara tree has a compact growth habit and subtle leaf variegation, with the cambium area of stems and bark in some trees being reddish in color. The fruit is attractive in salads due to its near crimson flesh containing a red pigment called lycopene, the same pigment found in the pink and red grapefruit. The flesh color develops well during warm weather, unlike the red pigmentation of the blood oranges which requires cool weather for development.

Variety:
AMBERSWEET
(USDA release, 1989)

Type and parentage:
Orange hybrid
[(Clementine X Orlando)
Midseason Orange]
Average diameter
(inches): 2 1/2 - 3 3/4
Seeds per fruit: 0 - 30
(seedless or nearly so,
if not cross pollinated)
Commercial harvest
season:
October - January

Ambersweet, a sweet orange hybrid released in 1989 by the USDA, has been classified as an orange for fresh fruit and processing purposes. The fruit is medium to large in size, slightly pear shaped, and has a medium orange rind color at maturity. The peel of very small developing Ambersweet fruit has a pubescent (fuzzy) appearance, a characteristic which disappears as fruit mature. Seed number will vary, being almost seedless in solid plantings and very seedy in mixed plantings. It is peeled more easily than other oranges and the juice color is good. Early information indicated it did not require cross pollination, but more recent observations indicate that cross pollination may increase yields. Trees are moderately cold hardy and fruit can usually be harvested prior to freeze events. Trees up to six years old in commercial plantings have produced relatively low yields of poor quality fruit. However, the outlook for both production and quality improves with age especially where irrigation and nutrition programs are carefully managed. Trees propagated on Swingle citrumelo rootstock tend to produce higher yields than those on Cleopatra mandarin. Rootstock selection, precise cultural management and innovative marketing are some of the keys to the success of Ambersweet as a commercial variety.

Variety:
HAMLIN

Type and parentage:
Sweet orange
Average diameter (inches): 2 3/4 - 3
Seeds per fruit: 0 - 6
Commercial harvest season:
October - January

Hamlin arose as a chance seedling in a grove planted in 1879 near Deland by Judge Issac Stone and later purchased by A. C. Hamlin. Hamlin is the most widely grown early-season sweet orange variety in Florida. Harvest is usually possible before the onset of freeze events in Florida. Because of its high yields, Hamlin is the most productive orange on the basis of pounds of solids per acre even though solids per box are lower than for midseason varieties and Valencia. Small fruit size can be a problem for the fresh fruit market, particularly during heavy crop years. Its juice color is relatively poor, particularly on vigorous rootstocks, so the juice is blended with that of other varieties to meet color standards. Fruit of this variety can develop good natural external color break early in the season, but usually requires degreening. Wood dieback and splitting and creasing of fruit can sometimes result in heavy fruit drop later in the season.

Variety:
PARSON BROWN

Type and parentage:
Sweet orange
Average diameter (inches): 2 1/2 - 2 3/4
Seeds per fruit: 10 - 20
Commercial harvest season:
October - January

Parson Brown originated as a chance seedling at the home of Reverend N. L. Brown near Webster, Florida in 1856. Original plantings of this variety on sour orange rootstock on heavy hammock soils produced good crops of relatively high quality fruit as would Hamlin under similar conditions. While Parson Brown may be harvested slightly earlier than the Hamlin, it is not as productive but has slightly better juice color. Its fruit quality is generally mediocre and seediness makes it less desirable than Hamlin for the fresh market.

Variety:
PINEAPPLE

Type and parentage:
Sweet orange
Average diameter
(inches):   2 3/4 - 3
Seeds per fruit: 15 - 25
Commercial harvest
season:
December - February

Pineapple sweet orange originated from a seedling planted near Citra, Florida by the Reverend J. B. Owens around 1860. Pineapple orange has been the leading midseason variety with good internal quality and juice color. Pineapple fruit is seedy, well-sized and the peel develops a reddish-orange color. This variety is alternate bearing. However, once the fruit matures, it is very susceptible to preharvest drop particularly during heavy crop years, a condition which may be alleviated with growth regulator sprays. Long-term yields may be lower than those of Hamlin and Valencia, and it is very susceptible to blight and freezes, particularly when heavily cropped. The peel is subject to creasing and to another stress-related condition known as 'pitting'. While the excellent quality juice is favored by processors, the variety is no longer planted for many of the above-mentioned reasons.

Variety:
SUNSTAR
(USDA release, 1987)

Type and parentage: Sweet orange

Average diameter (inches): 2 1/2 - 3

Seeds per fruit: 6 - 20

Commercial harvest season:

December - March

Sunstar, released by the USDA in 1987, is a midseason orange selection that ripens about the same time as Pineapple, about two weeks earlier than Midsweet and Gardner and has a slightly darker juice color than Hamlin. Its productivity equals that of Hamlin and exceeds that of Pineapple. Trees are more cold hardy and subject to less preharvest fruit drop than Pineapple.

Variety:
MIDSWEET
(USDA release, 1987)

Type and parentage:
Sweet orange
Average diameter (inches): 2 3/4 - 3
Seeds per fruit: 6 - 24
Commercial harvest season:
January - March

Midsweet, a midseason orange selection released by the USDA in 1987, ripens later than Pineapple, about the same time as Gardner and holds well on the tree. Midsweet produces approximately the same quantity of fruit and pounds of solids per acre as Hamlin and juice color score is higher than that of Hamlin. Trees are more cold hardy, less susceptible to preharvest fruit drop and are considered a good replacement for Pineapple.

Variety:
GARDNER
(USDA release, 1987)

Type and parentage:
Sweet orange
Average diameter (inches): 2 1/2 - 3
Seeds per fruit: 8 - 24
Commercial harvest season:
January - March

Gardner is a midseason orange selection released by the USDA in 1987. It produces about the same amount of fruit and pound solids as Pineapple or Valencia and has slightly darker juice color than Hamlin, Sunstar and Midsweet. Fruit ripens later than Pineapple and at the same time as Midsweet. Fruit rind creasing is less than that of Pineapple and trees are more cold hardy than Pineapple.

Variety:
VALENCIA

Type and parentage:
Sweet orange
Average diameter (inches): 2 3/4 - 3
Seeds per fruit: 0 - 6
Commercial harvest season:
March - June

    Valencia most likely originated in Spain or Portugal and was imported into Florida about 1870. Valencia has a wide range of climatic adaptability and is the leading sweet orange variety both in Florida and the world. Fruit production is basically lower than that of early varieties. It accounts for about 50 percent of the Florida crop, where it is the principal variety for processing. Its excellent internal quality, including juice color, makes it desirable for both processed and fresh markets. Valencia is the only variety carrying the old and new crops on the tree after bloom. As a late variety it is unlikely to be harvested before the occurrence of a freeze. Trees tend toward alternate bearing especially if fruit is harvested late in the season. While fruit store well on the tree, regreening of the peel can occur late in the season. Trees of Valencia are quite susceptible to the fungus *Colletotrichum acutatum*, the cause of postbloom fruit drop (PFD). A number of selections are available for planting including the Rohde Red Valencia with its superior peel and internal flesh color. Some early selections of Rohde Red Valencia have performed poorly in commercial plantings.

ROHDE RED

15

Variety:
SANGUINELLI

Type and parentage:
Blood orange
Average diameter (inches) 2 -2 3/4
Seeds per fruit: 4 -10
Commercial harvest season:
February - April

This variety, not commercially grown in Florida, is called a blood orange because of the red flecks of pigment (anthocyanins) found in the flesh of the fruit late in the season. As the development of pigment is dependent upon extended periods of cool weather, fruit of this variety is more likely to attain the greatest amount of red pigment in the northern citrus areas of Florida. Internal fruit quality of this variety is somewhat similar to that of a midseason type.

GRAPEFRUIT

Grapefruit probably originated in the West Indies, and was first reported growing in Barbados about 1750. It was brought to Florida in 1823 by Count Odette Phillipe near Safety Harbor and is grown throughout tropical and subtropical regions of the world. Florida produces more grapefruit than any other state in the U.S. or country in the world. Fruit is oblate to round and seed number and flesh color vary depending on the variety. The grapefruit peel is moderately thick and bright yellow at maturity, though some of the pigmented varieties containing lycopene will exhibit a pink or red blush. As internal quality is better in warmer climates, grapefruit matures later in the northern citrus-producing areas of the state. Trees can be vigorous in nature due to rootstock. Over-fertilization or severe pruning tend to produce fruit which are pear-shaped or sheep-nosed. Grapefruit mature as early as October and may be harvested as late as May. Grapefruit juice has long had a small but loyal following; current growth of the market is in the premium, pink or Ruby Red variety of ready-to-serve juices. Fruit held late in the marketing season is subject to drop, seed sprouting and rind aging. Applications of plant growth regulators can be applied to reduce drop and delay rind aging. When grapefruit is grown for the fresh market special attention must be paid to the control of scab, greasy spot rind blotch and melanose to maintain external grade.

| Star Ruby | Rio Red | Flame | Ray Ruby | Ruby |

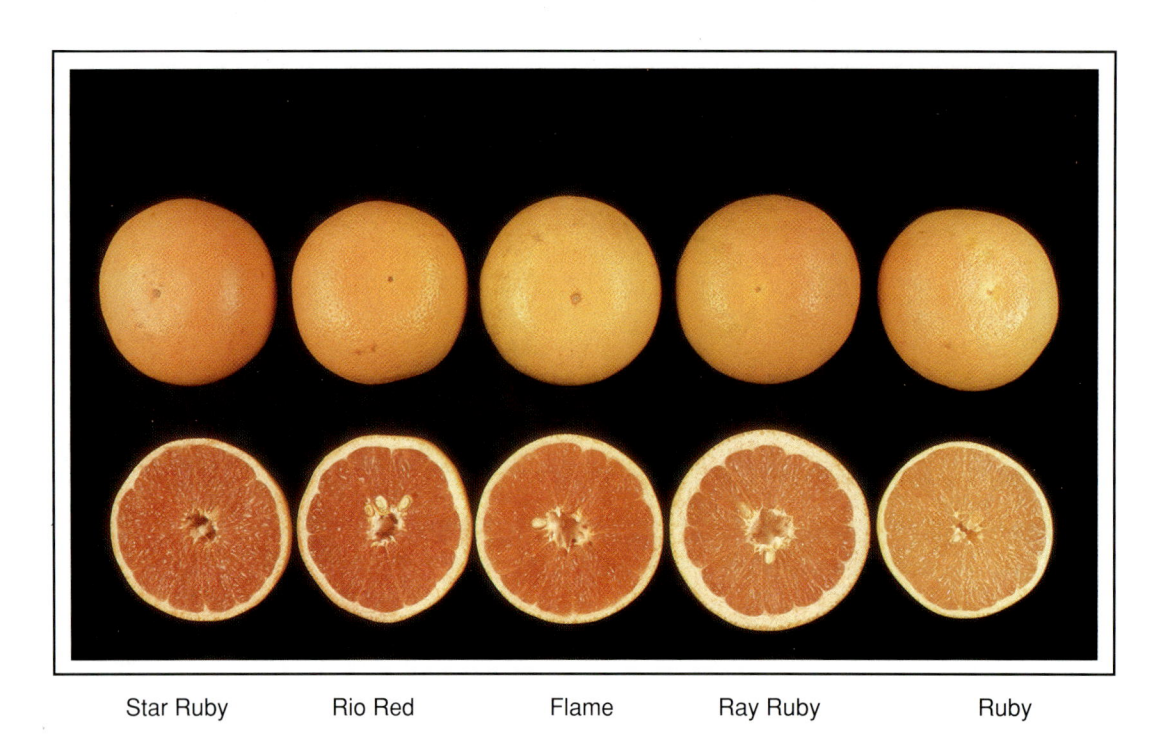

Peel and Internal Color Comparisons of Red Grapefruit Varieties

Variety:
DUNCAN

Type and parentage:
Grapefruit
Average diameter (inches): 3 1/2 - 5
Seeds per fruit: 30 - 70
Commercial harvest season:
December - May

Duncan is one of the oldest grapefruit varieties grown in Florida and is believed to have originated as a seedling about 1830 near Safety Harbor. While the seediness of Duncan excludes it from the fresh fruit market, it is considered to be one of the best-tasting grapefruit varieties. It has always been popular for sectionizing, but labor cost now prohibits this utilization and most fruit are processed.

Variety:
MARSH

Type and parentage: Grapefruit
Average diameter (inches): 3 1/2 - 4 1/2
Seeds per fruit: 0 - 6
Commercial harvest season:
November - May

Marsh most likely was discovered as a chance seedling near Lakeland, Florida around 1860. It was the most widely grown grapefruit variety in Florida until recently when the pink and red selections gained prominence. The fruit is commercially seedless, has a pale yellow flesh, and a large open cavity in the center. It retains its popularity in the domestic and export fresh markets and remains a major grapefruit variety for both fresh utilization and processing. Nearly all of the pigmented grapefruit varieties grown today have originated from a series of Marsh mutations.

Variety:
FOSTER

Type and parentage:
Grapefruit
Average diameter
(inches): 3 1/2 - 5
Seeds per fruit: 30 - 50
Commercial harvest
season:
November - March

Foster, the first pink-fleshed variety on record, occurred as a limb sport of Walters grapefruit. As with Duncan, the extreme seediness of Foster makes it unpopular for fresh fruit consumption and only a very small acreage remains. A pink blush color can be detected in the peel of the fruit offering exceptional early season external quality.

Variety:
**THOMPSON
(PINK MARSH)**

Type and parentage: Grapefruit
Average diameter (inches): 3 3/4 - 4 1/2
Seeds per fruit: 0 - 6
Commercial harvest season:
December - May

Thompson or Pink Marsh arose as a limb sport of Marsh grapefruit near Oneco, Florida in the mid 1920s, and only a very small acreage remains. Thompson was the first internally pigmented seedless grapefruit variety but has been replaced by the more popular and deeply pigmented varieties. Fruit characteristics are essentially identical to that of Marsh with the exception of the pink flesh color which fades as the season progresses, with no pink blush being detected in the peel.

Variety:
REDBLUSH
(RUBY, RUBY RED)

Type and parentage: Grapefruit
Average diameter (inches): 3 1/2 - 4 1/2
Seeds per fruit: 0 - 6
Commercial harvest season:
November - May

Redblush, originating as a limb sport of Thompson in Texas about 1930, is the most widely grown seedless colored grapefruit in Florida, particularly in the Indian River area. The peel exhibits a pink blush and the flesh is pink to pale red in color, fading somewhat later in the season. It is used for the fresh fruit market, but its popularity is declining in favor of the more recently introduced red selections.

Variety:
RAY RUBY

Type and parentage: Grapefruit
Average diameter (inches):   3 3/4 - 4 1/2
Seeds per fruit: 0 - 6
Commercial harvest season:
November - May

Ray Ruby was first observed growing in a Texas grove of Redblush grapefruit. The fruit peel has a darker pink blush than Redblush (Ruby) but less than that of Star Ruby, and about the same as that of Flame and Rio Red. The flesh color is darker than that of Redblush and slightly lighter than that of Flame, especially late in the season.

Ray Ruby's internal color holds better late in the season, and a greater peel blush is usually noted for this variety than for Redblush. Segments have darker colored flesh near edges than in the center and taste and juice content are similar to that of other varieties except Star Ruby.

24

Variety:
FLAME
(USDA release, 1987)

Type and parentage:
Grapefruit
Average diameter (inches): 3 1/2 - 4
Seeds per fruit: 0 - 6
Commercial harvest season:
November - May

Flame grapefruit, a selection from Henderson seedlings planted in 1973 at the USDA Whitmore Foundation Farm near Leesburg, is commercially seedless. The fruit peel has a darker pink blush than Redblush but less than that of Ray Ruby, Rio Red and Star Ruby. The flesh color is significantly darker than that of Redblush, slightly darker than Ray Ruby, and slightly lighter than Rio Red. Trees exhibit heavy internal canopy bearing in clusters. Trees exhibit mild chlorotic patterns on the foliage (of genetic origin) and show sensitivity to certain herbicides.

Variety:
RIO RED

Type and parentage:
Grapefruit
Average diameter (inches):   3 3/4 - 4 1/2
Seeds per fruit: 0 - 6
Commercial harvest season:
November - May

Rio Red originated as a bud sport mutation of an unreleased Texas seedling selection, first produced by seed irradiation. The fruit peel has a darker pink blush than Redblush, less than that of Star Ruby and about the same as that of Ray Ruby and Flame. The flesh color is about the same as Flame, significantly darker than that of Redblush, slightly darker than Ray Ruby, but not as dark as Star Ruby. Some reports indicate that as with Ray Ruby, flesh color is not always uniform across segments. Trees are less sensitive to herbicides than Star Ruby and Flame.

Variety:
STAR RUBY

Type and parentage:
Grapefruit
Average diameter (inches): 3 1/2 - 4
Seeds per fruit: 0 - 6
Commercial harvest season:
December - May

This variety originated from irradiated seed of Hudson (a seedy pink-fleshed variety) at Texas A&M University, and was released in 1970. Short internodes, profuse branching and a compact habit of tree growth characterize this variety. The fruit is characterized by a dark pink blush in the peel and very intense deep red flesh. The fruit is seedless and smaller than Ray Ruby and Rio Red. Juice is deeply colored even late in the season. Foliage often shows blotchy chlorotic areas (a genetic trait), and the trees are very sensitive to some herbicides. Trees are less cold hardy than other grapefruit varieties and are more susceptible to Phytophthora foot rot. The yield in some locations is erratic and less than desired, and efforts to increase production have been unsuccessful. Propagation of Star Ruby in Florida peaked in the 1984-85 season, declining due to cultural problems and the release of other red grapefruit selections.

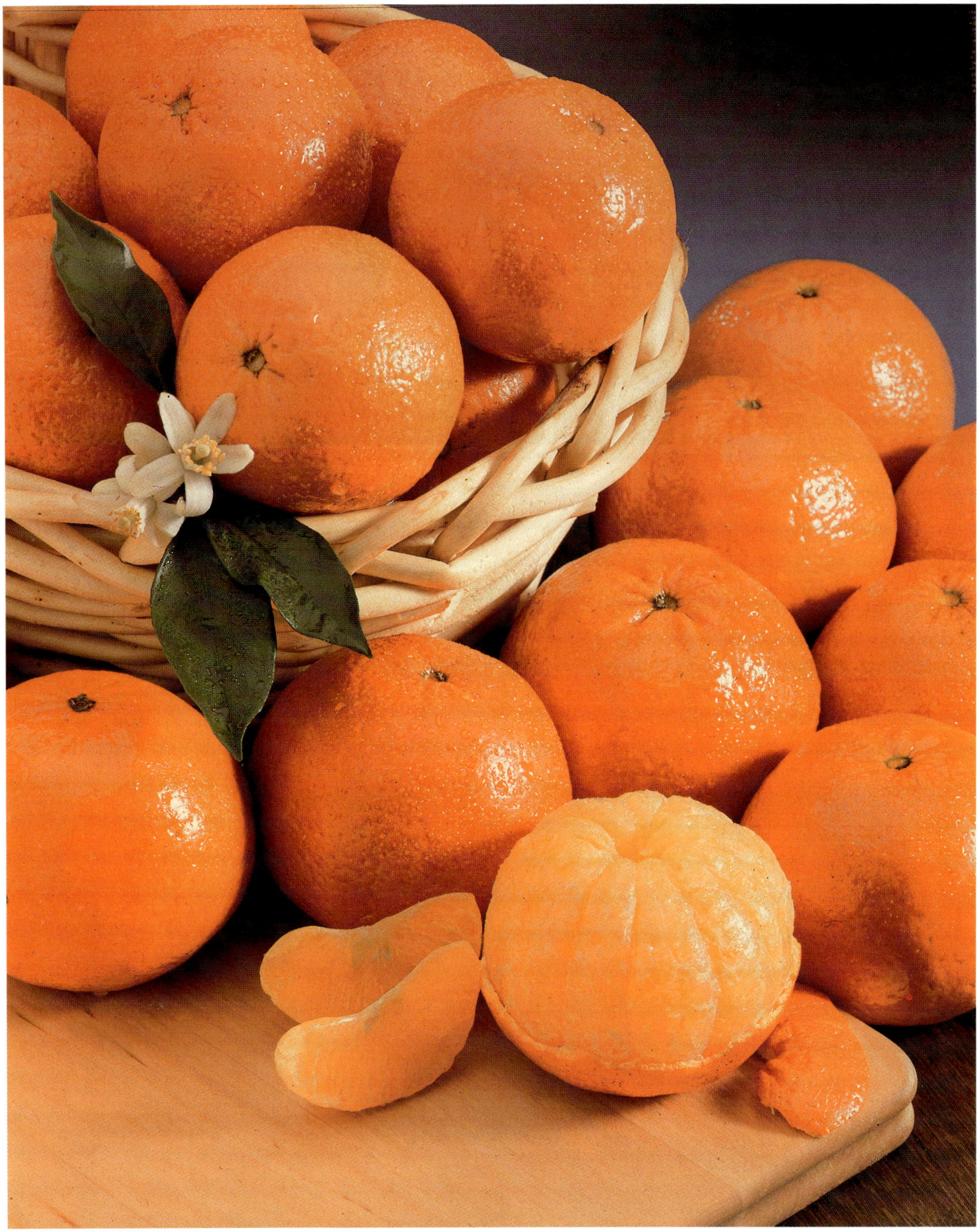

TANGERINES AND TANGERINE HYBRIDS

Tangerines (sometimes referred to as mandarins) and their hybrids represent the most diverse group of citrus varieties grown in Florida. In general, they are more compact and cold hardy trees than sweet orange or grapefruit. Wood of some varieties is brittle making trees prone to limb breakage. The fruit are more highly pigmented internally and externally than oranges, and they are characterized typically as more aromatic and frequently sweeter than oranges. The fruit of some of these varieties can be peeled more easily than sweet oranges. Many of the varieties grown in Florida are self- and cross-incompatible and require pollination by other compatible varieties to set acceptable crops. However, with adequate cross-pollination the fruit become more seedy and less acceptable in the market place. There are some naturally seedless varieties, such as Clementine and the Satsumas, but these are not widely grown because they are not so well adapted to Florida conditions. Fruit of tangerines and their hybrids are produced primarily for the fresh fruit market, though some volume is juiced and blended with orange juice to improve juice color.

Variety:
MINNEOLA
(USDA release, 1930)

Type and parentage:
Tangelo
(Duncan X Dancy)
Average diameter
(inches): 3 - 3 1/2
Seeds per fruit: 7 - 12
Commercial harvest season:
December - February

The Minneola tangelo, sometimes referred to as Honeybell, is a Duncan grapefruit x Dancy tangerine hybrid released by the USDA. Minneola fruit is characterized by a stem-end neck which tends to make the fruit appear pear-shaped, an appearance which has given rise to the name Honeybell in the gift fruit trade. The peel is relatively thin and smooth, and tends to adhere to the internal fruit surface making it difficult to peel. Minneola is not strongly self-compatible and will produce greater yields when interplanted with suitable pollenizers including Temple, Sunburst or Fallglo. Fruit produced on trees in solid plantings of Minneola are likely to be commercially seedless, while trees in mixed plantings will typically be seedy due to cross-pollination. Fruit maturing in the December-February period has found ready acceptance in the gift fruit trade because of its handsome appearance and unique eating quality. Minneola trees are quite vigorous and will develop into large trees if given adequate room. Minneola is susceptible to greasy spot, scab and particularly Alternaria brown spot fungus diseases which necessitates a vigorous spray program during spring and early summer to prevent damage to leaves and fruit and ultimately yield loss.

Variety:
ORLANDO
(USDA release, 1930)

Type and parentage:
Tangelo
(Duncan X Dancy)
Average diameter (inches): 2 3/4 - 3
Seeds per fruit: 0 - 35
Commercial harvest season:
November - January

The Orlando tangelo is a Dancy tangerine x Duncan grapefruit hybrid made by the USDA in 1911. The foliage of Orlando exhibits a distinctive cupped appearance. The color and texture is similar to that of sweet orange and the rind adheres firmly making it difficult to peel. Seed number per fruit will vary depending on extent of cross-pollination with a low of zero in solid blocks to as many as 35 seeds in blocks pollinated with Temple, Robinson or Sunburst. Orlando is used as a pollenizer for many citrus hybrids. Orlando tangelos have been observed to require higher rates of nitrogen than most other varieties particularly in the fall and winter when premature yellowing of leaves, a condition referred to as winter chlorosis, is frequently observed. Orlando tangelos are also susceptible to Alternaria and scab fungus diseases.

Variety:
NOVA
(USDA release, 1964)

Type and parentage:
Tangelo
(Clementine X Orlando)
Average diameter
(inches): 2 3/4 - 3
Seeds per fruit: 1 - 30
Commercial harvest
season:
November - December

Nova is a hybrid of Clementine tangerine x Orlando tangelo similar to that which produced the Robinson, Osceola and Lee released in 1964 by the USDA. Nova is self-incompatible and must be planted with pollenizers including Temple, Orlando, Lee and Sunburst to set an adequate fruit crop. While Orlando is acceptable as a pollen source, Nova will not produce adequate pollen as a pollenizer for Orlando. The fruit has a tendency to mature before external color break and prematurely dry out when trees are grown on the more vigorous rootstocks. The fruit will not retain the necessary quality for shipping if subjected to extended degreening.

Variety:
ROBINSON
(USDA release, 1959)

Type and parentage:
Tangerine
(Clementine X Orlando)
Average diameter (inches): 2 1/2 - 2 3/4
Seeds per fruit: 1 - 20
Commercial harvest season:
October - December

    The Robinson, officially released in 1959, is one of three citrus hybrids originating from a Clementine x Orlando cross made by the USDA in 1942. As the fruit is 3/4 tangerine and 1/4 grapefruit from the Orlando parent, it looks like and is marketed as a tangerine. Seed content varies greatly, being dependent upon cross-pollination with Temple, Orlando, Sunburst or Lee. The fruit tends to dry out prematurely on trees on vigorous rootstocks, particularly if held beyond maturity. Splitting of the thin-skinned fruit is often a problem. Harvest prior to color break followed by a long degreening period can lead to postharvest problems, particularly anthracnose caused by the fungus *Colletotrichum gloeosporioides*. The Robinson tree growth habit and brittle wood with a large crop load can result in limb breakage. It is susceptible to twig and limb dieback and dead wood must be pruned out. The Robinson scion is also susceptible to Phytophthora foot rot infection. It has become less popular commercially due to the various cultural problems and market preference for other varieties.

Variety:
SUNBURST
(USDA release, 1979)

Type and parentage:
Tangerine hybrid
(Robinson X Osceola)
Average diameter
(inches): 2 1/2 - 3
Seeds per fruit: 1 - 20
Commercial harvest season:
November - December

Sunburst is a cross between the two citrus hybrids, Robinson and Osceola, made in 1961 and released by the USDA in 1979. This variety is self-incompatible and must be cross pollinated with Temple, Orlando, Nova, or Minneola. In years of heavy fruit set, thinning must be considered as a means of increasing fruit size and minimizing limb breakage. Sunburst trees may exhibit severe brownish-colored leaf and stem blistering from excessive mite feeding and other stress conditions and must be carefully monitored as the condition can result in premature leaf drop and wood dieback. Fruit should exhibit at least partial color break prior to harvest to reduce degreening time and postharvest problems. This variety has been widely planted and currently enjoys good market acceptance.

Variety:
DANCY

Type and parentage: Tangerine
Average diameter (inches): 2 1/4 - 2 1/2
Seeds per fruit: 6 - 20
Commercial harvest season:
December - January

Dancy is one of the oldest tangerine varieties grown in Florida, originating in 1867 from a seed of Moragne tangerine in the grove of Colonel F. L. Dancy of Orange Mills. Dancy, a very popular variety over the years, is no longer widely planted due to numerous production problems including small fruit size and poor holding quality. The fruit is of excellent quality and the easily removed rind has deep reddish-orange color at maturity. The fruit is usually clipped at harvest to prevent plugging of peel. Dancy is self-compatible and therefore requires no pollenizers to enhance productivity. This variety is alternate bearing, producing large crops of small fruit one year followed by small crops of large fruit, a problem which has created a marketing challenge. Mechanical and chemical thinning of large crops can reduce alternate bearing and alleviate the small fruit size problem. Both the fruit and foliage are susceptible to Alternaria brown spot which can result in defoliation, fruit drop and crop reduction.

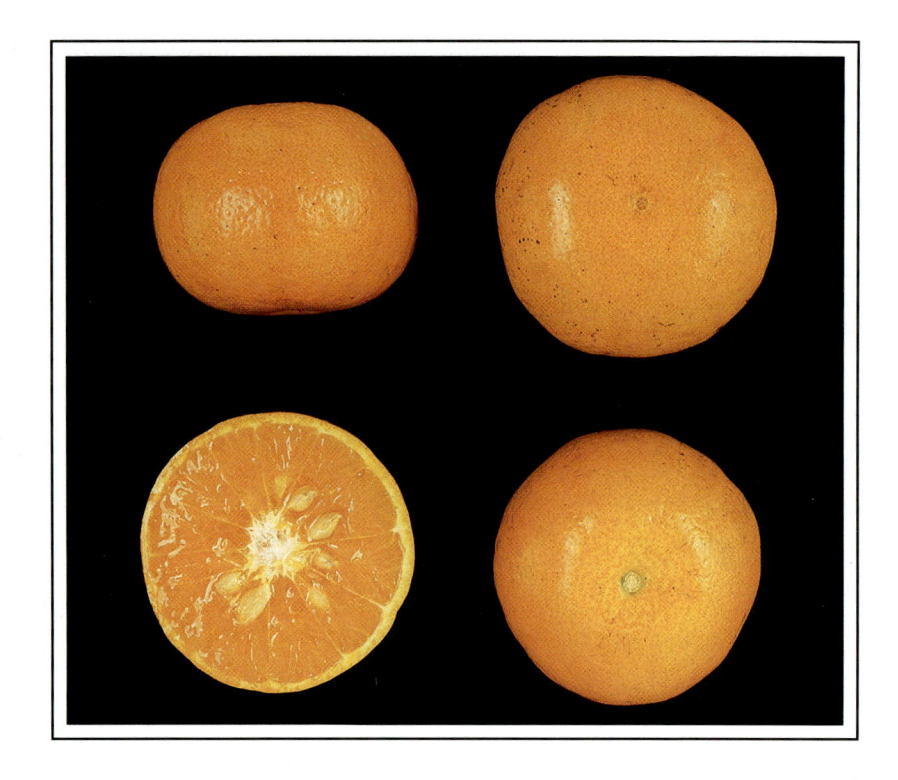

**Variety:
MURCOTT
(HONEY TANGERINE)**

Type and parentage: Probably a hybrid of a tangerine and sweet orange
Average diameter (inches): 2 3/4
Seeds per fruit: 10 - 20
Commercial harvest season:
January - March

The origin of the Murcott, marketed as Honey Tangerine, is unknown but it is most likely a tangor which is a cross between a tangerine and sweet orange. This variety may have originated in a USDA planting around 1916. The fruit is exceptionally sweet but more difficult to peel than a tangerine. The peel is yellowish-orange and the flesh is a deep orange at maturity. Murcotts exhibit alternate bearing, and in years of very large crops trees are prone to breakage, collapse and even death. Manual, mechanical or chemical fruit thinning is essential to reduce crop load. In years of large crops the nutritional inputs particularly nitrogen and potassium, must also be increased due to the greater nutrient demand extending into the fall of the year. The current rootstock of choice is Cleopatra mandarin as compatibility problems have occurred with trifoliate orange, and its hybrids Carrizo citrange and Swingle citrumelo. Fruit and leaves are susceptible to scab and Alternaria brown spot. Fruit are highly susceptible to wind scar and sunburn because of the exposed nature of the fruit borne on the outside of the tree canopy.

Variety:
TEMPLE
(TEMPLE ORANGE)

Type and parentage:
Probably a hybrid of tangerine and sweet orange
Average diameter (inches): 2 3/4 - 3
Seeds per fruit: 15 - 20
Commercial harvest season:
January - March

Temple originated in Jamaica in the late 1800s, and although frequently referred to as an orange, there is evidence to suggest it is actually a tangerine x orange hybrid or tangor. It is one of the more cold-tender tangerine hybrids. The fruit rind, with a pebbly or somewhat rough surface, is fairly thick and relatively easy to remove. Temple, which produces an abundance of pollen, is often used as a pollenizer in blocks of self-incompatible varieties. Temple fruit and leaves are very susceptible to scab fungus disease which can severely disfigure the fruit and leaves. The foliage is also highly susceptible to aphid damage.

Variety:
OSCEOLA
(USDA release, 1959)

Type and parentage:
Citrus hybrid
(Clementine X Orlando)
Average diameter
(inches): 2 1/4 - 2 3/4
Seeds per fruit: 15 - 25
Commercial harvest season:
October - November

Osceola is one of three citrus hybrids arising from a Clementine mandarin x Orlando tangelo cross by the USDA, officially released in 1959. Although highly colored, it was never widely accepted commercially primarily because of its mediocre flavor.

Cross-pollination with Orlando tangelo or Temple is essential to insure good productivity and fruit size. The fruit and leaves are susceptible to scab fungus disease and fruit should be harvested with clippers to avoid plugging of the peel.

Variety:
LEE
(USDA release, 1959)

Type and parentage:
Citrus hybrid
(Clementine X Orlando)
Average diameter
(inches): 2 3/4 - 3
Seeds per fruit: 10 - 25
Commercial harvest season:
November - December

    Lee is one of three citrus hybrids arising from a Clementine mandarin x Orlando tangelo cross by the USDA, officially released in 1959. Cross-pollination was apparently thought to be unnecessary, but yields have been enhanced when pollenizers such as Orlando, Page or Temple were nearby. Peel color develops slowly after fruit maturity and does not peak until the fruit is nearly over-mature, necessitating extensive degreening which often results in postharvest problems.

Variety:
FALLGLO
(USDA release, 1987)

Type and parentage:
Citrus hybrid
(Bower X Temple)
Average diameter
(inches): 2 3/4 - 3 1/4
Seeds per fruit: 20 - 40
Commercial harvest
season:
October - November

    The citrus hybrid Fallglo is a cross between Bower tangerine and Temple made by the USDA in 1962 and released in 1987. The variety is characterized by its large fruit size, narrow leaf shape and pale foliage color. The tree bears consistently and does not require cross-pollination but has been used as a pollenizer. The variety often suffers from a twig and limb dieback, particularly during early tree development. Fruit should be harvested with good color-break to minimize degreening which can result in poor holding quality.

Variety:
PAGE
(USDA release, 1963)

Type and parentage:
Citrus hybrid
 (Minneola X Clementine)
Average diameter
(inches): 2 - 2 1/2
Seeds per fruit: 0 - 25
Commercial harvest season:
October - February

Page is a hybrid of Minneola tangelo x Clementine mandarin released in 1963 by the USDA. A high quality fruit, Page lost market acceptance mainly due to small fruit size, although there is some improvement with cross-pollination. The fruit and foliage are susceptible to scab. It should remain an option for homeowners.

Variety:
SATSUMA

Type and parentage:
Mandarin
Average diameter (inches): 2 1/4 - 2 1/2
Seeds per fruit: 0 - 6
Commercial harvest season:
September - November

Satsuma, originating in Japan, is one of the earliest maturing mandarin types. Several varieties are grown in the U.S., with Owari being the most common. Trees have a characteristic open habit of growth with less foliage than other varieties. The tree is more cold hardy than most other citrus varieties enabling it to be grown in northern areas of the state. There is limited production in Florida as optimum fruit quality requires cool fall and winter weather. The looseness of the skin and low tolerance to degreening contribute to its poor shipping quality, and fruit must be clipped to avoid plugging.

Variety:
PONKAN

Type and parentage:
Mandarin
Average diameter (inches): 2 3/4 - 3 1/4
Seeds per fruit: 3 - 7
Commercial harvest season:
December - January

Ponkan is regarded as an excellent quality, easily peeled dooryard mandarin with many unnamed budlines in Florida that vary in fruit quality characteristics. The trees tend toward alternate bearing. Fruit packs and ships poorly and it must be clipped as the peel plugs easily during harvesting.

ACID CITRUS FRUIT

The lack of cold hardiness has limited the commercial plantings of acid citrus fruit, particularly lemons and limes, to warmer locations in the southern portion of Florida. Thorniness, tree vigor and multiple cropping are problems commercially.

Variety:
TAHITI
(PERSIAN)

Type and parentage: Lime
Average diameter (inches): 1 3/4 - 2 1/2
Seeds per fruit: 0 - 1
Commercial harvest season:
June - September

Tahiti is not a true lime, but is probably a product of hybridization resulting in seedlessness. Trees are vigorous, thorny and highly susceptible to cold injury, limiting its culture to south Florida. For a lime, the fruit is large and is spot picked while green based on size and juice content. The seedless fruit, when fully ripe, is pale yellow. Fruit is susceptible to a physiological stylar end breakdown at maturity often associated with harvesting and handling procedures.

Variety:
KEY LIME
(MEXICAN)

Type and parentage:
Lime
Average diameter (inches): 1 1/4 - 1 3/4
Seeds per fruit: 3 - 8
Commercial harvest season:
Everbearing

Although once grown in the Florida keys, Key limes are not now grown commercially in the U.S. Key lime trees flower repeatedly and consequently have fruit in various stages of development on the tree. While most trees have many thorns, thornless selections are available. Trees of this variety are very susceptible to cold. The fruit is smaller and seedier than the Tahiti lime and the sharper flavor is unlike that of the larger Persian or Tahiti lime. The fruit peel turns yellow when fully ripe. Key limes are prized for their flavor for pies and other culinary uses.

Variety:
MEYER

Type and parentage:
Probably a
lemon hybrid
Average diameter
(inches): 2 1/2 - 3
Seeds per fruit: 10
Commercial harvest
season:
November - March

Meyer lemon hybrid was a popular selection for dooryard plantings. The tree, with a characteristic low spreading growth habit and few thorns, is the most cold hardy lemon variety. The fruit is relatively large for a lemon and often has a nipple at the blossom end. It has lower acid content than other lemon varieties and is not suitable for lemon oil production. Some selections of Meyer contain severe strains of citrus tristeza virus (CTV) which may not cause problems for Meyer trees, but present a potential source of infection for susceptible scion rootstock combinations. It has, therefore, been recommended that existing Meyer trees be destroyed and replanted with certified disease-free budlines to protect commercial citrus production.

Variety:
BEARSS
(SICILIAN)

Type and parentage:
Lemon
Average diameter (inches): 2 1/2 - 3 1/2
Seeds per fruit: 1 - 6
Commercial harvest season:
July - December

Bearss lemon, the commercial lemon variety of Florida, was all but eliminated during the freezes of the 1980s. Trees are very vigorous, thorny and prone to continuous vegetative regrowth requiring frequent pruning to control tree size. Trees are also very sensitive to cold and scab fungus disease. The fruit, while still green, is picked for size and later harvested for processing and peel oil recovery.

MISCELLANEOUS

Many of these lesser known varieties, as well as many citrus relatives are located at the Florida Citrus Arboretum maintained by Florida Department of Agriculture and Consumer Services (FDACS) Budwood Registration Bureau located in Winter Haven, Florida.

Variety:
CALAMONDIN

Type and parentage:
Probably a hybrid of a sour mandarin and some kumquat
Average diameter (inches): 1 1/4 - 1 1/2
Seeds per fruit: 3 - 5
Commercial harvest season:
November - April

    The Calamondin is a cold hardy tree which is quite showy when the fruit is mature. As such it is widely used in landscaping or as ornamental plants in large containers. Plants from rooted cuttings are marketed as miniature oranges for use as winter houseplants. The Calamondin fruit with its distinctive acid flavor is used to flavor drinks and for marmalade and jelly.

Variety:
NAGAMI
(KUMQUAT)

Type and parentage:
Kumquat
Average diameter
oblong (inches):
1 1/4 - 1 3/4 X 3/4 - 1 1/4
Seeds per fruit: 0 - 3
Commercial harvest
season:
November - April

The Nagami or oblong kumquat differs from Meiwa with its longer fruit, more acid taste, and brighter orange color. The tree is more vigorous and attains a greater size than Meiwa. Nagami trees are used in home and commercial landscaping and are quite cold hardy. Careful rootstock selection is necessary for successful tree growth of all kumquat varieties. It makes excellent preserves.

Variety:
**MEIWA
(KUMQUAT)**

Type and parentage: Probably a chance hybrid
Average diameter - globose
(inches): 1 - 1 1/2
Seeds per fruit: 3 - 5
Commercial harvest season:
November - April

The trees, of small to medium size with dark green foliage and compact shape, are used in home and commercial landscaping and are quite cold hardy. Meiwa is a large round kumquat used for preserves, candied fruit and is one of the best for eating out of hand. Careful rootstock selection is necessary for successful tree growth of all kumquat varieties.

Variety:
TAVARES
(LIMEQUAT)

Type and parentage:
Citrus hybrid
(East Indian lime X
oblong kumquat)
Average diameter -
oblong (inches):
1 3/4 - 2 X
1 1/4 - 1 1/2
Seeds per fruit: 2 - 5
Season of maturity:
November - March

    Limequat trees, more cold hardy than limes but less than kumquats, are also popular for home landscaping. Fruit of Tavares limequat are more characteristic of the oblong kumquat than the lime parent. This acid fruit may substitute for lime as a condiment.

Variety:
HIRADO BUNTAN SEEDLING

Type and parentage: Pummelo
Average diameter (inches): 5 - 7
Seeds per fruit: 50 +
Commercial harvest season:
November - February

The Pummelo is a large oblate to round fruit that resembles grapefruit but the inside texture is firmer and less juicy than grapefruit. The tree is less cold hardy than most grapefruit and has very large leaves, stems and flowers. There is a strong tendency toward alternate bearing. This fruit is more of a dessert or salad fruit and is not eaten with a spoon as is grapefruit. The taste is sweeter and less acid than grapefruit. This variety has pink flesh and yellow, smooth peel and originated as a seedling from the Japanese white-fleshed Hirado Buntan variety. The peel is thick but sensitive to mechanical injury and does not store well after harvest compared to other Pummelos. This variety has the best taste of the pink-fleshed varieties tested in Florida.

56

Variety:
NAKON

Type and parentage:
Pummelo
Average diameter (inches): 5 - 7
Seeds per fruit: 50
Commercial harvest season:
December - February

The Nakon originated as seeds from the Nakon Chaisri of Thailand and is similar to the Wainwright variety in south Florida. The tree is less cold hardy than grapefruit and has large leaves and stems. The fruit is elongated or pear-shaped and the peel has a pebbly texture due to prominent oil glands. The peel is thick and fruit store well after harvest. The flesh is greenish in color and has a firm texture. The fruit is not as juicy as other pummelos and is prone to dry out or granulate. It is good for sectionizing and is used in salads and desserts.

NOTES

NOTES

NOTES